Teen N
363.738
Terp, Gail
Climate and environmental inju
33410018300410   01-16-2023

EXPLORING SOCIAL INJUSTICE
CLIMATE AND ENVIRONMENTAL INJUSTICE

by Gail Terp

BrightPoint Press

San Diego, CA

© 2023 BrightPoint Press
an imprint of ReferencePoint Press, Inc.
Printed in the United States

For more information, contact:
BrightPoint Press
PO Box 27779
San Diego, CA 92198
www.BrightPointPress.com

ALL RIGHTS RESERVED.

No part of this work covered by the copyright hereon may be reproduced or used in any form or by any means—graphic, electronic, or mechanical, including photocopying, recording, taping, web distribution, or information storage retrieval systems—without the written permission of the publisher.

Content Consultant: Katya Forsyth; Creative Director, Global Center for Climate Justice

LIBRARY OF CONGRESS CATALOGING-IN-PUBLICATION DATA

Name: Terp, Gail, author.
Title: Climate and environmental injustice / by Gail Terp.
Description: San Diego, CA: BrightPoint Press, Inc., 2023 | Series: Exploring social injustice | Includes bibliographical references and index. | Audience: Grades 10–12
Identifiers: ISBN: 9781678203948 (hardcover) | ISBN: 9781678203955 (eBook)
The complete Library of Congress record is available at www.loc.gov.

CONTENTS

AT A GLANCE   4

INTRODUCTION   6
   WHAT IS CLIMATE AND
   ENVIRONMENTAL INJUSTICE?

CHAPTER ONE   12
   HISTORY OF CLIMATE AND
   ENVIRONMENTAL INJUSTICE

CHAPTER TWO   26
   CLIMATE AND ENVIRONMENTAL
   INJUSTICE TODAY

CHAPTER THREE   34
   KEY MOMENTS IN FIGHTING CLIMATE
   AND ENVIRONMENTAL INJUSTICE

CHAPTER FOUR   46
   ENDING CLIMATE AND
   ENVIRONMENTAL INJUSTICE

Glossary   58
Source Notes   59
For Further Research   60
Index   62
Image Credits   63
About the Author   64

AT A GLANCE

- Environmental injustice occurs when some groups are harmed more than others by environmental issues.

- Climate injustice happens when the effects of climate change hurt some groups more than others.

- As the climate warms, oceans are warming, too. This causes sea levels to rise.

- Social justice issues such as redlining have led to environmental and climate injustice.

- The goal of the Paris Agreement of 2015 was to reduce greenhouse gas emissions.

- City officials in Flint, Michigan, changed the source of the city's water. This change caused Flint's water to contain lead, which caused learning problems for many children in Flint.

- Research has shown that most environmental groups are led by white people. Activists are working to change that.

- A meeting called COP26 was held in 2021. Its goal was to take action against climate change.

INTRODUCTION

WHAT IS CLIMATE AND ENVIRONMENTAL INJUSTICE?

In 1978, North Carolina looked for a place to build a toxic waste site. Toxic waste is chemical waste that is harmful. It comes from many sources. These include factories, power plants, and hospitals. Toxic waste must be disposed of properly. If it is not, it pollutes the soil, air, and water. It can make people sick.

The state chose Warren County for the waste site. Warren County was poor. More than half of its citizens were Black. They did not want the waste site. Warren County citizens believed the state would not have

An unclean water supply can have a serious effect on wildlife and public health.

Government leaders may choose to start a landfill near a poor community.

chosen a rich, white county for the site. They were outraged. Hundreds marched and protested. Ken Ferruccio was one of the protest leaders. He said, "These folks believe that they're fighting for their lives, more so now than ever."[1]

The waste site was ultimately placed in Warren County. But news of the protest spread. It inspired Black leaders. They wanted to prove that waste sites were placed more often in neighborhoods of color than in white ones. In the 1980s, researchers did prove it. Decision makers often choose areas with fewer resources for toxic waste sites. People in these areas often don't have the resources to fight back. This is an example of environmental injustice. This injustice happens when environmental issues harm some groups more than others. Issues include air, water, and soil pollution. Environmental injustice builds on existing social injustices such as systemic racism and poverty.

Climate change is a long-term change in weather patterns. Weather is becoming more extreme. Extreme weather often hurts some groups more than others. When it does, it's called climate injustice.

Environmental and climate injustice are linked with environmental racism. This occurs when decisions about the environment harm people of color more than other groups. Toxic waste is dumped

Climate change causes an increase in extreme weather, including floods.

near their homes. Factories pollute their air and water.

Leaders in government make decisions. Studying patterns in the effects of these decisions can help reveal climate and environmental injustice. The more people see the problems, the more they can work to fix them.

1
HISTORY OF CLIMATE AND ENVIRONMENTAL INJUSTICE

Climate and environmental injustice are not new. Their roots come from a long history of segregated housing in the United States. One contributing factor was the National Housing Act. The US government passed this act in 1934. The program helped poor people buy homes. It did so by guaranteeing home loans. That meant banks could safely lend money to home buyers. However, the act kept loans from going to people of color. The government drew maps. Neighborhoods of color were

Systemic racism can make it difficult for people from communities of color to receive bank loans.

outlined in red. Loans were not guaranteed in those areas. This practice was called redlining.

The National Housing Act had a big impact. Most people of color weren't able to buy homes. They had to rent.

THE INDUSTRIAL REVOLUTION

The environmental justice movement began in the 1970s. But environmental injustice has existed for much longer. Before the 1700s, most people lived on farms and in towns. The Industrial Revolution changed things. People built factories in cities. Lots of people moved to cities for work. Cities did not have enough housing. Families were jammed into too-small buildings. The factories created pollution. It made people sick. Wealthy people left the cities for cleaner country air.

They had little control over the condition of their buildings. It was more difficult to move away from things such as toxic waste sites. The Fair Housing Act of 1968 banned housing discrimination. But years of discrimination would not end quickly. Many neighborhoods remained segregated.

In the late 1970s, the battle in Warren County was not the only waste site fight. A toxic waste site was planned for a Black neighborhood in Houston, Texas. Researchers looked into Houston's waste sites. All five were in Black neighborhoods. Hundreds of people protested.

People continue to fight for environmental justice today.

The Houston and Warren County protests brought attention to environmental injustice. In 1987, the United Church of Christ (UCC) did new research. Researchers looked at where toxic waste sites were built. They wrote a report. It proved that toxic waste sites in the United States were most often placed in neighborhoods of color. The report became a key tool for the environmental justice movement.

ENVIRONMENTAL LEADERSHIP SUMMIT

In 1991, a meeting was held in Washington, DC. It was called the First National People of Color Environmental Leadership Summit. Hundreds of Black, Hispanic and Latino, Asian, and Native American people attended. Dana Alston was one of the summit's leaders. She wrote, "For people of color, environmental issues are not just a matter of preserving ancient forests or defending whales. While the importance of saving endangered species is recognized, it is also clear that adults and children living in communities of color are endangered species too."[2]

Summit attendees shared stories of environmental injustice they had faced. They also worked to find solutions. Summit delegates wrote the *17 Principles of Environmental Justice*. The *Principles* set several goals. One was to protect the right to clean air, land, water, and food. The *Principles* called for the production

The EPA headquarters are located in Washington, DC.

of toxic and hazardous materials to stop. The end goal was not equal pollution for everyone. It was no pollution at all.

In 1992, the Environmental Protection Agency (EPA) published a report called *Environmental Equity: Reducing Risk for All Communities*. The report shared research from the EPA's Environmental Equity Workgroup. The report said that people of color and low-income groups lived with higher environmental risk. They were more exposed to air pollution. They more often lived near toxic waste sites. The EPA set up the Office of Environmental Equity later that year. The name later changed to the Office of Environmental Justice. President Bill Clinton declared environmental justice a national priority in 1994. Government recognition was important. But more progress would be needed to solve these problems.

Oil, coal, and natural gas are all fossil fuels. Gasoline is made from oil.

CLIMATE INJUSTICE

Climate change is caused by an increase of greenhouse gases. These gases are found in Earth's atmosphere. Most come from burning **fossil fuels**. Greenhouse gases are making Earth warmer. This can lead to big

Many of the poor and majority Black communities on the east side of New Orleans received little government aid following Hurricane Katrina.

weather problems. These include floods, droughts, and hurricanes. Such weather events happen more often than they did in the past. They are often more deadly.

These weather problems do not affect all people equally. For example, Hurricane Katrina struck the US Gulf Coast in 2005.

It was one of the deadliest hurricanes in US history. More than 1,800 people died. Katrina hit New Orleans, Louisiana, hard. All parts of the city were flooded. But low-lying neighborhoods flooded the most. Redlining was used in New Orleans in the 1930s. The redlined areas were in the lowest spots. Most Black residents were limited to living in those areas. When Katrina hit, Black homeowners were three times more likely than white homeowners to experience flooding.

After the storm, it was time to rebuild. Black neighborhoods got less help. It took longer to fix their homes. White neighborhoods got help first. Greta Gladney works for justice in New Orleans. She said, "It was the people with means who got back to the city first, and they were the ones making the decisions."[3] Nearly twenty years later, some historically Black neighborhoods still had not fully recovered.

Climate change continues to cause extreme weather worldwide. Countries have looked for ways to fight it. In 2015, they gathered to write the Paris Agreement. The agreement's goal was to reduce greenhouse gases. Countries had to show

President Barack Obama was one of many world leaders who spoke in support of the Paris Agreement in 2015.

how they would reduce their greenhouse gas **emissions**. Richer countries agreed to help poorer ones finance the work. Most of the world's nations signed. The agreement went into effect in 2016.

2
CLIMATE AND ENVIRONMENTAL INJUSTICE TODAY

Climate injustice continues to the present. Climate change can cause severe heat waves. These long periods of hot weather can lead to drought. Droughts are long periods of little or no precipitation. Lack of water affects people, animals, and crops.

Drought is a growing problem in the southwestern United States. It greatly affects many Native American peoples living there. When white settlers came to the Southwest, they forced Native American peoples onto reservations. Reservations are smaller than a people's traditional

Lake Mead provides water to many people in the Southwest. Lower water levels due to drought especially affect poor communities.

lands. Reservations often have less rain. Today, drought has dried up community wells. More than one-third of the Navajo Nation lives without running water. Native Americans can request government

DROUGHT AND THE NAVAJO NATION

The Navajo Nation has been hit hard by drought. Navajo Nation president Jonathan Nez said, "Our people are right when they say that water is life. We see what is happening all over the Southwest. We are getting less and less moisture every year, our lakes and ponds are drying up, and our wells are depleting."

Quoted in Katrina Machetta, "Navajo Nation Continues to Experience Drought," *Navajo-Hopi Observer*, October 12, 2021. www.nhonews.com.

assistance to recover from climate disasters. But they are less likely to receive aid than non-Native communities are.

Coastal areas have a different problem. As the climate warms, so does the ocean. Warming water expands. It takes up more space. Warming water and air melt ice sheets and glaciers. Expanding water and melting ice cause **sea levels** to rise. As sea levels rise, land is washed away. Floods increase. More than one-third of the world's population lives near coasts. Some experts predict 200 million people worldwide will have to move by 2100.

CURRENT ENVIRONMENTAL INJUSTICE

Environmental injustice also continues. In 2007, UCC released an update to its groundbreaking 1987 report. It studied the twenty years since the first report was published. Researchers found that racial disparities in exposure to waste were higher than originally reported. Neighborhoods with toxic waste sites were 56 percent people of color. Neighborhoods without waste sites were 30 percent people of color.

But the report was about more than just where people lived. It also found that

Protesters in New York City gathered in 2020 to fight for climate justice.

government officials did not respond in quick and effective ways. The EPA is in charge of enforcing environmental rules. These rules cover hazardous waste, toxic chemicals, and more. When the EPA finds an environmental problem, it is supposed to find the source. Then the people responsible must fix and clean up the problem. The EPA lists environmental justice as a goal. But its actions don't always reflect this goal.

In 2016, the US Commission on Civil Rights published a report. It examined how well the EPA was meeting its responsibilities. It found that the EPA still struggled to help communities of color. And the EPA did not take action until forced to do so. The report found the EPA had delays in enforcement. When the EPA was sued, it would settle the lawsuit. But it wouldn't dig deeper to find the source of the problem.

The report suggested bringing on additional staff to help the EPA fulfill environmental justice goals.

CANCER ALLEY

An 85-mile (140-km) stretch of land in Louisiana has more than 140 **petrochemical** plants. The plants cause air pollution. A 2015 EPA study found that this area had the highest risk of cancer caused by air pollution in the United States. The area has become known as Cancer Alley. Sharon Lavigne is an activist from St. James, Louisiana. She said, "We are boxed in from all sides by plants, tank farms, and noisy railroad tracks. . . . They're killing us. And that is why I am fighting."

Quoted in Antonia Juhasz, "Louisiana's 'Cancer Alley' Is Getting Even More Toxic—But Residents Are Fighting Back," Rolling Stone, October 30, 2019. www.rollingstone.com.

3
KEY MOMENTS IN FIGHTING CLIMATE AND ENVIRONMENTAL INJUSTICE

A few events have brought major attention to the environmental justice movement. One is the Dakota Access Pipeline (DAPL) protests. DAPL is 1,172 miles (1,886 km) long. It moves crude oil from North Dakota to Illinois. The pipeline runs near land of the Standing Rock Sioux

Tribe. It poses a risk to the water used by the tribe.

Oil began flowing through the line in 2017. The Standing Rock Tribe fought DAPL from the start. They held protests.

The Dakota Access Pipeline travels under the Missouri River, threatening the water supply of the Standing Rock Sioux.

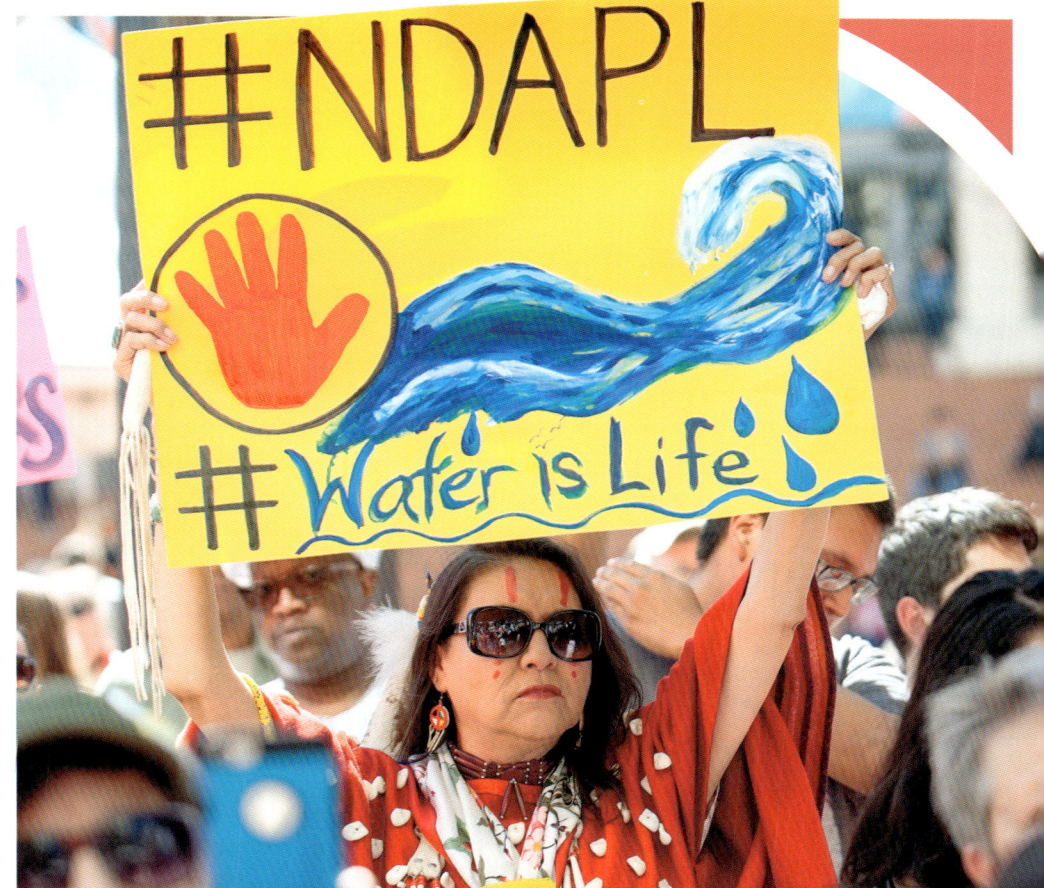

People from around the country joined. The pipeline company fought back. Its crews attacked with dogs and pepper spray. They destroyed ancestral burial sites. In 2020, a judge ordered DAPL to close down. However, the pipeline company appealed. A new judge ruled that the line could stay open.

LINE 3 PIPELINE

Line 3 is a tar sands pipeline. Tar sands are a mix of sand, clay, water, and thick oil. The line opened in 2021. It crosses Minnesota's watersheds and tribal lands. Tribal groups and others fought the pipeline. They held protests. They also challenged the pipeline legally.

FLINT WATER CRISIS

Flint is a majority Black city in Michigan. In 2014, city officials stopped piping Flint's water from Detroit. Instead, the city piped its water from the Flint River. Soon after this switch, Flint residents complained about the water. It was brown and smelled bad. People were getting sick.

 EPA tests showed that the water had high levels of lead. The river water was high in chloride. As it flowed through the lead pipes, lead went into the water. Lead can cause brain damage. It's especially dangerous for children. City officials knew about the high lead levels. Even so, they told the public not to worry. Brad Wurfel worked with the Michigan Department of Environmental Quality. He said, "Let me start here—anyone who is concerned about lead in the drinking water in Flint can relax."[4]

A study done in 2015 found troubling news. Flint children had high lead levels in their blood. The city then switched back to Detroit water. In 2017, the city began replacing the lead water pipes. But the damage was already done. Many children developed learning problems. Even when the EPA declared Flint's water safe to

The National Guard delivered bottled water to Flint, Michigan, in 2016.

drink, people did not trust it. "When you tell us that the water is safe but it really wasn't, that relationship between leadership and the community is still damaged," Flint pastor Todd Womack said. "That just layers the historical trauma that has presented itself in our community."[5]

The river water led to other problems. An outbreak of Legionnaires' disease killed at least twelve people. In 2021, nine Michigan officials were charged with crimes for their handling of the water crisis.

HURRICANE IDA

Hurricane Ida hit New York City in 2021. The storm caused floods that killed thirteen people. Eleven of them were living in basement apartments. Nearly all were of Asian descent. Most of the flooded

A New York building inspector examines a flooded basement apartment after Hurricane Ida.

apartments were illegal. They did not meet building codes. People had complained about the poor conditions. Inspectors made two calls to each complaint site. However, they weren't able to get in. They then closed each case.

The deaths caused by Hurricane Ida flooding raised concerns. Bill de Blasio was the mayor of New York City. He proposed a plan to prevent more flooding. The plan called for an early flood warning system. It also contained ways to reduce future flood risk. But locals were doubtful that things would change.

Lina Lee is the executive director of Communities Resist. This is a housing justice nonprofit organization. She said, "It is only when there is a tragedy like the victims of Hurricane Ida that the city pays attention to the plights of Asian American tenants."[6]

Diverse perspectives are needed to fight climate injustice.

DIVERSITY AND FUNDING

The group Green 2.0 began in 2014. It works to increase **diversity** in environmental groups. Green 2.0 sponsored research about diversity in the groups' leaders. *The State of Diversity in*

Environmental justice groups rallied in Philadelphia, Pennsylvania, in 2018.

Environmental Organizations shared the findings. Since the 1991 Environmental Summit, little diversity progress had been made. The report showed an increase of female leaders in the groups. But it also showed there were still few racially diverse leaders. Green 2.0 has released more research about how to increase diversity.

Funding is another issue. White-led environmental groups still receive the most funding. In 2021, Democratic lawmakers proposed legislation to fund environmental justice groups. Other activists encouraged existing groups to focus on environmental justice.

Demos is a research and advocacy group. Heather McGhee is a senior fellow there. She said, "It's essential to have anti-racism baked into the goals . . . because both political racism and environmental racism are drivers of our excess pollution and climate denialism."[7]

4
ENDING CLIMATE AND ENVIRONMENTAL INJUSTICE

Many people and groups fight for climate and environmental justice. Some are people in power. Some are ordinary people. They want to be part of making change happen. All are key to the fight.

The Green New Deal was first sent to Congress in 2019. At that time, Congress did not pass it. The bill was brought to Congress again in 2021. It had several goals to be met over ten years. The bill would

The Sunrise Movement is a youth-led climate justice organization. Members gathered in Washington, DC, to support the Green New Deal.

reduce greenhouse gases and toxins in the air. It would fund clean-energy projects. The projects would create millions of new jobs. The bill would also improve the nation's **infrastructure**. It would provide health care and safe housing. The bill also included plans to clean up toxic waste.

The bill had both support and criticism. Supporters believed cleaner air and water would improve people's health. This would save on health care costs. More jobs would lead to people having more to spend. This would help the economy. Those who were against the bill thought it was too

President Joe Biden has pushed for climate justice. But some critics believe he has not done enough.

expensive. They also thought the bill would give the government too much power over energy companies.

NATIONAL AND GLOBAL STEPS

In 2021, President Joe Biden issued Executive Order 14008. Its aim was to fight the climate crisis. The order had many parts. One would form the White House Environmental Justice Interagency Council. Its goal was to work for environmental justice. It would look at past and present problems. The council would work with the EPA.

A meeting called COP26 was held in 2021. *COP* stands for Conference of the Parties. It was the twenty-sixth year that COP had been held. Its goal was to take action against climate change. The first COP was held in 1995. The Paris Agreement was written at COP21 in 2015.

ND-GAIN INDEX

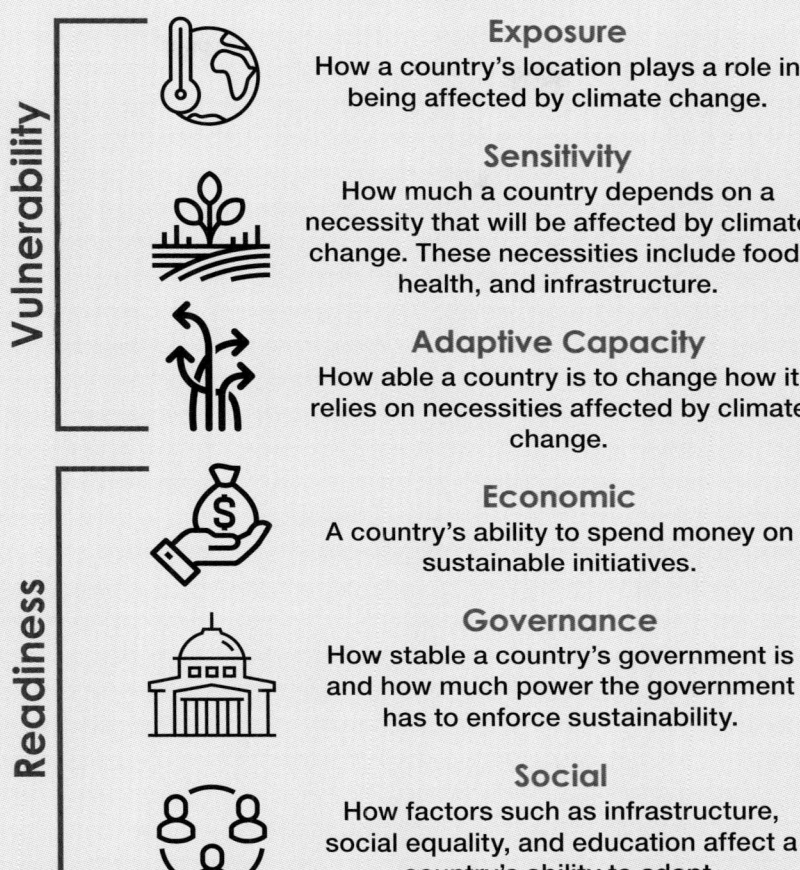

Vulnerability

Exposure
How a country's location plays a role in being affected by climate change.

Sensitivity
How much a country depends on a necessity that will be affected by climate change. These necessities include food, health, and infrastructure.

Adaptive Capacity
How able a country is to change how it relies on necessities affected by climate change.

Readiness

Economic
A country's ability to spend money on sustainable initiatives.

Governance
How stable a country's government is and how much power the government has to enforce sustainability.

Social
How factors such as infrastructure, social equality, and education affect a country's ability to adapt.

Source: "Country Index," ND-GAIN, July 2021, https://gain.nd.edu.

The Notre Dame Global Adaptation Initiative (ND-GAIN) country index examines how easily different countries will be able to adapt to climate change. Factors such as government stability and poverty levels play a role. The criteria show how factors beyond weather play a role in climate injustice.

COP26 took place in Glasgow, Scotland. People from nearly 200 nations were there. They **negotiated** various goals for two weeks. One goal was to cut emissions from burning coal, oil, and gas. Another goal was to slow the rise of global temperature. It said that the global temperature should rise no more than 2.7 degrees Fahrenheit (1.5°C). If it were to rise higher, there would be more deadly heat waves, droughts, and crop failures. Sea levels would rise, causing more floods. All of these dangers would increase climate injustice.

People in Glasgow gathered outside COP26. They wanted to make sure world leaders were serious about fighting for climate justice.

While COP26 met, the COP26 Coalition met outside. These people wanted to make sure COP26 prioritized climate justice. They came to demand that world leaders act. They didn't want just words. They wanted action. In Glasgow, more than 100,000 people marched. Protests were also held worldwide. People of all ages and backgrounds came to these marches.

Vanessa Nakate was at the march in Glasgow. She is from Uganda. She said, "Leaders rarely have the courage to lead. It takes citizens, people like you and me,

PRIVATE JETS TO COP26

Many of the world leaders involved in COP26 flew to Glasgow on private jets. Critics thought the jets were a bad choice. Jets use a huge amount of fuel. Matt Finch is the UK policy manager of the Transport and Environment campaign group. He said, "It is difficult to avoid the hypocrisy of using [private jets] while claiming to be fighting climate change."

Quoted in Ollie A. Williams, "118 Private Jets Take Leaders to COP26 Climate Summit Burning over 1,000 Tons of CO2," *Forbes*, November 5, 2021. www.forbes.com.

to rise up and demand action. And when we do that in great enough numbers, our leaders will move."[8]

CLIMATE AND ENVIRONMENTAL ORGANIZATIONS

There are many groups that fight climate and environmental injustice. One is the **Indigenous** Environmental Network (IEN). IEN unites Indigenous peoples worldwide. It works to protect sacred sites, as well as land, water, and air. IEN shares Indigenous knowledge with others. It works to change unfair laws.

350.org is also a global group. It works for climate justice. It works with other climate justice groups to get their voices heard. 350.org also helps train people to be climate leaders. In 2020, 350.org worked with several groups. In New York City, its work led to a ban on all fossil fuel projects. In Brazil, it helped groups stop an open-pit coal mine. It also helped stop a coal project in Kenya.

Earthjustice is a legal group. Its lawyers practice environmental law. Earthjustice uses the legal system to protect Earth and its people. When it believes a company has broken the law, it brings the case to court. Earthjustice works for its clients for free.

The fight for climate and environmental justice is far from over. Lasting change requires work done on all levels. People and communities, justice groups, and world leaders are key to the fight.

GRETA THUNBERG AND FRIDAYS FOR FUTURE

Greta Thunberg lives in Sweden. In 2018, she began to fight climate change. On Fridays, she would skip school. Holding a large sign, she stood outside the Swedish Parliament. The sign said, School Strike for Climate. Her protest hit the news. Youth of the world began to protest. The Fridays for Future movement is now worldwide.

GLOSSARY

diversity

having people of different races or cultures in a group or organization

emissions

substances released into the air, such as exhaust fumes from cars

fossil fuels

fuels such as oil, coal, or natural gas that are formed in the earth from dead plants or animals

Indigenous

having to do with the earliest known residents of an area

infrastructure

the basic equipment and structures a country needs to function properly, such as bridges and roads

negotiated

held a discussion between parties to come to an agreement or solution

petrochemical

a chemical that is made from petroleum or natural gas

sea levels

the average height, between high and low tide, of the sea's surface

SOURCE NOTES

INTRODUCTION: WHAT IS CLIMATE AND ENVIRONMENTAL INJUSTICE?

1. Quoted in Matt Reimann, "The EPA Chose This County for a Toxic Dump Because Its Residents Were 'Few, Black, and Poor,'" *Timeline*, April 2, 2017. https://timeline.com.

CHAPTER ONE: HISTORY OF CLIMATE AND ENVIRONMENTAL INJUSTICE

2. Dana Alston, "The Summit: Transforming a Movement," *Race, Poverty, & the Environment*, spring 2010. www.reimaginerpe.org.

3. Quoted in Gary Rivlin, "Why the Lower Ninth Ward Looks Like the Hurricane Just Hit," *The Nation*, August 13, 2015. www.thenation.com.

CHAPTER THREE: KEY MOMENTS IN FIGHTING CLIMATE AND ENVIRONMENTAL INJUSTICE

4. Quoted in Lindsey Smith, "Leaked Memo Shows Concerns About Lead," *Michigan Radio*, July 13, 2015. www.michiganradio.org.

5. Quoted in Erin Einhorn, "Scars from Flint's Water Crisis Shake City's Faith in COVID Vaccine," *NBC News*, January 12, 2021. www.nbcnews.com.

6. Quoted in Kimmy Yam and Sakshi Venkatraman, "Ida's Forgotten Victims," *NBC News,* October 18, 2021. www.nbcnews.com.

7. Quoted in Somini Sengupta, "Black Environmentalists Talk About Climate and Anti-Racism," *New York Times*, June 3, 2020. www.nytimes.com.

CHAPTER FOUR: ENDING CLIMATE AND ENVIRONMENTAL INJUSTICE

8. Quoted in "COP26: Thousands March for Glasgow's Biggest Protest," *BBC News*, November 6, 2021. www.bbc.com.

FOR FURTHER RESEARCH

BOOKS

Paul Douglas, *A Kid's Guide to Saving the Planet: It's Not Hopeless and We're Not Helpless*. Minneapolis, MN: Beaming Books, 2022.

Naomi Klein and Rebecca Stefoff, *How to Change Everything: The Young Human's Guide to Protecting the Planet and Each Other*. New York: Atheneum Books for Young Readers, 2021.

Julie Knutson, *Do the Work! Clean Water and Sanitation*. Ann Arbor, MI: Cherry Lake Publishing, 2022.

INTERNET SOURCES

"Do You Have a Right to Clean Environment?" *Youth Civil Rights Academy*, n.d. https://youthcivilrights.org.

Carolyn Kormann, "The Teen-Agers Fighting for Climate Justice," *New Yorker*, July 22, 2018. www.newyorker.com.

"Principles of Environmental Justice," *EJNet.org*, April 6, 1996. https://www.ejnet.org/ej/principles.html.

WEBSITES

NAACP: Environmental & Climate Justice
https://naacp.org/know-issues/environmental-climate-justice

The NAACP provides information about the issues involved in environmental and climate justice. It also provides information on ways to take action.

EPA: Environmental Justice
www.epa.gov/environmentaljustice

This website of the Environmental Protection Agency has information about environmental justice. It also has links for deeper research.

Fridays for Future
https://fridaysforfuture.org

This website describes what Fridays for Future does and provides steps for taking action about climate change.

INDEX

Alston, Dana, 17–18

Biden, Joe, 49–50

Cancer Alley, 33
climate change, 10, 21–22, 24–25, 26, 50–52, 54, 57
Clinton, Bill, 20
COP26, 50–54

Dakota Access Pipeline (DAPL), 34–36
de Blasio, Bill, 42
drought, 22, 26–28, 52

Earthjustice, 56–57
Environmental Justice Interagency Council, 50
Environmental Protection Agency (EPA), 19–20, 31–33, 37, 50

Ferruccio, Ken, 8
Finch, Matt, 54
First National People of Color Environmental Leadership Summit, 17–19
Flint, Michigan, 37–40
flooding, 22–23, 29, 40–42, 52

Gladney, Greta, 24
Green 2.0, 43–44
Green New Deal, 47–49
greenhouse gas emissions, 21, 24–25, 47–48

Hurricane Ida, 40–42
Hurricane Katrina, 22–24

Indigenous Environmental Network (IEN), 55
Industrial Revolution, 14

Lavigne, Sharon, 33
Lee, Lina, 42
Line 3, 36

McGhee, Heather, 45

Nakate, Vanessa, 54
Navajo Nation, 28
Nez, Jonathan, 28

Paris Agreement, 24–25, 50
Principles of Environmental Justice, 18–19

redlining, 12–15, 23

350.org, 56
Thunberg, Greta, 57
toxic waste sites, 6–11, 15–17, 20, 30, 48

United Church of Christ (UCC), 16–17, 30

Warren County, 7–9, 15–16
Womack, Todd, 39–40
Wurfel, Brad, 38

IMAGE CREDITS

Cover: © Jim West/Alamy
5: © Leo Patrizi/iStockphoto
7: © Jon Shore/Shutterstock Images
8: © Vchal/Shutterstock Images
11: © Roschetzky Photography/Shutterstock Images
13: © Fizkes/Shutterstock Images
16: © Phil Pasquini/Shutterstock Images
19: © Visual Field/iStockphoto
21: © Scarc/Shutterstock Images
22: © J. Norman Reid/Shutterstock Images
25: © Frederic Legran - COMEO/Shutterstock Images
27: © Wonderlust Pics Travel/Shutterstock Images
31: © Ron Adar/Shutterstock Images
35: © Diego G. Diaz/Shutterstock Images
39: © Linda Parton/Shutterstock Images
41: © Ron Adar/Shutterstock Images
43: © Cory Seamer/Shutterstock Images
44: © Rachael Warriner/Shutterstock Images
47: © Rachael Warriner/Shutterstock Images
49: © Biksu Tong/Shutterstock Images
51 (top): © Telman Bagirov/Shutterstock Images
51 (second): © Kuroksta/Shutterstock Images
51 (third): © GZP Design/Shutterstock Images
51 (fourth): © Davooda/Shutterstock Images
51 (fifth): © Alexandr III/Shutterstock Images
51 (bottom): © Davooda/Shutterstock Images
53: © Bruno Mameli//Shutterstock Images

ABOUT THE AUTHOR

Gail Terp is the author of more than seventy nonfiction books for children. A retired elementary teacher, she now enjoys writing about all sorts of topics, such as people, science, history, and the natural world. She also likes to search out fun and quirky books to recommend to her family and friends. When not reading and writing, she walks around looking for interesting stuff to write about.